Luana Pinheiro

Innovation to Reduce Air Conditioning Use

Luana Pinheiro

Innovation to Reduce Air Conditioning Use

Exploring Prototype Materials for Sustainable Cooling

ScienciaScripts

Imprint

Cover image: www.ingimage.com

This book is a translation from the original published under ISBN 978-620-6-76113-6.

Publisher:
Sciencia Scripts
is a trademark of
Dodo Books Indian Ocean Ltd. and OmniScriptum S.R.L publishing group

120 High Road, East Finchley, London, N2 9ED, United Kingdom
Str. Armeneasca 28/1, office 1, Chisinau MD-2012, Republic of Moldova, Europe
Printed at: see last page
ISBN: 978-620-8-15731-9

PAGE ABOUT THE AUTHOR

Luana Ribeiro Pinheiro, Civil Engineer, was born in 1999 in the city of Osasco, in the state of São Paulo. She lived her entire childhood and adolescence with her family in her hometown.

In 2018, at the age of 18, he took his first big step by moving to another Brazilian state to pursue his dream of going to university. She moved to the state of Pernambuco where she began studying for a Bachelor's degree in Civil Engineering at the Federal Rural University of Pernambuco (UFRPE) in Cabo de Santo Agostinho. She met wonderful people and a totally different culture to the one she knew, she was in a city she hadn't even known existed. It was a great challenge, but a watershed. She learned the value of her independence and how much she can embrace and face challenges.

However, in 2019 he felt that his career should change course and decided to return to his home state. That's where she found herself and started living in the city of Caraguatatuba, in the state of São Paulo. It's been five years of challenges and persistence.

In 2023, he published his first academic article on prototypes of innovative materials that could replace the use of air conditioning and fans, entitled "Innovation to Reduce the Use of Air Conditioning: Exploring Prototype Materials for Sustainable Cooling".

In 2024, at the age of 24, he graduated in Civil Engineering from the Federal Institute of Education, Science and Technology of São Paulo (IFSP), at the Caraguatatuba campus.

DEDICATORY

I dedicate this work to those who are part of my foundation in life. To my mother Sandra and my father Lourival, who gave me all their support and allowed me to live through everything I've lived through to get where I am. And to my love Vinícius, who is my joy, my inspiration, my support and my future. This book is dedicated to you, with all my love and gratitude.

PREFACE

In her final year of training, in 2023, Luana published her first work called "Innovation to Reduce the Use of Air Conditioning: Exploring Prototype Materials for Sustainable Cooling", a scientific academic article on prototypes of innovative and sustainable materials within Civil Engineering that promise to reduce the use of or even replace air conditioning and fans. The published article opened the door for the topic to be developed into a book.

This work explores low-cost, self-sustainable materials that have the ability to cool environments through low or no energy consumption in order to reduce or replace the use of air conditioning and thus reduce carbon dioxide emissions.

ACKNOWLEDGMENTS

I would like to thank all my family and friends who showed me support and made the process easier. I would also like to thank the Federal Institute of Education, Science and Technology of São Paulo, Caraguatatuba Campus and all my teachers and tutors who gave me the proper guidance and quality training to make this work possible.

CONTENTS

I. INNOVATION AND SUSTAINABILITY IN CONSTRUCTION

Since the dawn of humanity, innovation has been crucial to its development and growth. It has the capacity to add to and replace more archaic methods, speeding up and simplifying the lives of those who adopt it. Innovation is of great importance to the continuous evolution of society. As it is an old concept, it has gone through several definitions, however, according to Law No. 10.973/2004, better known as the "Innovation Law", the word "innovation" can be defined as:

> "Introduction of novelty or improvement in the productive and social environment that results in new products, services or processes or that comprises the addition of new functionalities or characteristics to an existing product, service or process that may result in improvements and in an effective gain in quality or performance (BRASIL, 2004)."

Law No. 10.973/2004 makes it clear that innovation is not just a concept related to the creation of new products, services or processes, but also the addition of new functionalities that improve what already exists. Thus, the act of innovating can help develop new products at low cost and with short production times, important factors that help companies develop a competitive advantage over their competitors.

In the construction industry, technological innovation can improve time and cost factors. Setbacks and delays are common as a result of various factors, such as poor planning, rework, late deliveries, etc. which result in longer work execution times and higher costs. It is therefore interesting to adopt resources that can reduce these factors. One of the ways that can help reduce construction time and costs is the use of technological innovation.

> "In a world where productivity and competitiveness are essential characteristics when it comes to conquering and maintaining the market, being in line with innovation and pursuing it is very important for the company to perform well and stand out in the construction

> industry, as technological innovation has been a factor that differentiates and stands out in terms of the search for process improvements." (POTT, Luana; EICH, Monique; ROJAS, Fernando, 2017)

Introducing improvements to existing materials or even developing new ones for the construction environment can bring significant benefits, including reducing processes, which can increase the company's profitability. According to Pott, Eich and Rojas (2017), "everyone is looking for quality and speed on construction sites, combined with cost savings. And all this is possible with technology, which develops new materials, new methods, new programs and more."

As well as improving construction time and costs, construction also needs to be concerned about the environmental impact generated. The construction industry is responsible for having major environmental impacts at all stages, from the removal of raw materials to the execution of the work. According to data from the Brazilian Association of Public Cleaning and Special Waste Companies (ABRELPE), the generation of Municipal Solid Waste (MSW) in 2022 was approximately 81.8 million tons, which corresponds to 224,000 tons per day (ABRELPE, 2022). In 2021, more than 48 million tons of Construction and Demolition Waste (CDW) were collected (ABRELPE, 2022). The large amount of construction waste negatively affects the environment and requires responsibility to repair the damage caused to nature. According to the United Nations Environment Programme (2022), "in fast-growing economies of developing countries, construction materials are set to dominate resource consumption, with associated greenhouse gas emissions expected to double by 2060."

"Sustainability applied to construction aims to reduce the problems caused by the archaic methods that still predominate in the sector." (CONCEIÇÃO;

SANTOS, p.427, 2021). The environmental issue has encouraged the creation of alternatives within the traditional construction system, in order to collaborate with sustainable development and to restore as much as possible the damage caused not only by the sector, but also by all the other areas and human actions that have led to environmental degradation.

One of the ways of reconciling construction with the environment can be called "sustainable construction".

> "Sustainable construction is a building system that makes conscious changes to its surroundings, meeting the building, housing and use needs of today's man, preserving the environment and natural resources, guaranteeing quality of life for current and future generations (Araújo, 2008 apud. ROTH; GARCIAS, p.124, 2009)."

According to Conceição and Santos (p.426, 2021), sustainable construction "reflects the result of a more balanced approach to the environment, from the preparation to the finishing of a constructed building. In this type of construction, environmental impacts are reduced and natural resources are used with maximum efficiency." The authors also add that there is a predominance of "rational use of water and energy efficiency with the intention of maintaining these resources for future generations."

To achieve these goals, actions such as the use of less aggressive materials and the rationalization of natural resources were necessary.

> "Some actions, such as: the use of less aggressive materials in general, reducing water and energy waste, the use of solar energy, actions aimed at improving air quality and internal space can make a big difference and have been implemented little by little in sustainable buildings (CONCEIÇÃO; SANTOS, p.427, 2021)."

There are several ways to make a building qualify as sustainable, one of which is to use materials that are different from those normally used. Reused or completely new materials that take a sustainable approach are welcome in the sector and can be called sustainable materials. "A so-called sustainable product has less polluting characteristics or has consumed less energy, water or natural resources in its production, when compared to conventional products" (CAIADO, 2014 apud. CONCEIÇÃO; SANTOS, 2021).

II. GLOBAL WARMING

It's no news that global warming is increasingly becoming one of the global problems that need attention. "Global warming is the abnormal increase in the average temperature of the planet recorded in recent years. This phenomenon is directly related to anthropogenic actions (man's activities)." (SÃO PAULO, 2023)

As this is a problem of global interest, the United Nations Environment Programme (UNEP) and the World Meteorological Organization (WMO) created the main international body responsible for monitoring and studying climate change in 1988: the Intergovernmental Panel on Climate Change (IPCC). In 2023, the body published its sixth assessment report "Climate Change 2023: Synthesis Report" on climate change in recent years. According to this report translated into Portuguese and made available by the Federal Government of Brazil, IPCC (p. 60, 2023), "human activities, mainly through greenhouse gas emissions, have unequivocally caused global warming, with the global surface temperature reaching a value 1.1°C higher between 2011-2020 than in the period 1850-1900"

Its main causes are related to anthropogenic activities and the release of gases into the atmosphere that cause the greenhouse effect. Actions such as burning, poor land use and excessive use of energy, mainly from large industries and the agricultural sector, contribute directly and cause the greatest impact.

> "Global greenhouse gas emissions continued to increase in the period 2010-2019, with historical and current unequal contributions from unsustainable energy use, land use and land-use change, lifestyles, and consumption and production patterns between regions, between and within countries, and between individuals" (IPCC, p.60, 2023).

The greenhouse effect mentioned by the agency corresponds to a layer of gases above the Earth's surface that keeps the planet warm. Naturally, the Sun emits solar rays into the Earth's interior which are reflected back into space, however, part of these rays are prevented by the layer, keeping the planet warm. This layer is made up of gases such as methane (CH_4), carbon dioxide (CO_2), nitrous oxide ($N_2 O$) and water vapor. However, with the increase in the emission of these gases, this layer blocks even more of the reflection of these rays into space, resulting in extreme heat and climate change in the interior of the planet. The following image (Figure 1) illustrates how the greenhouse effect works.

Figure 1 - How the Greenhouse Effect works

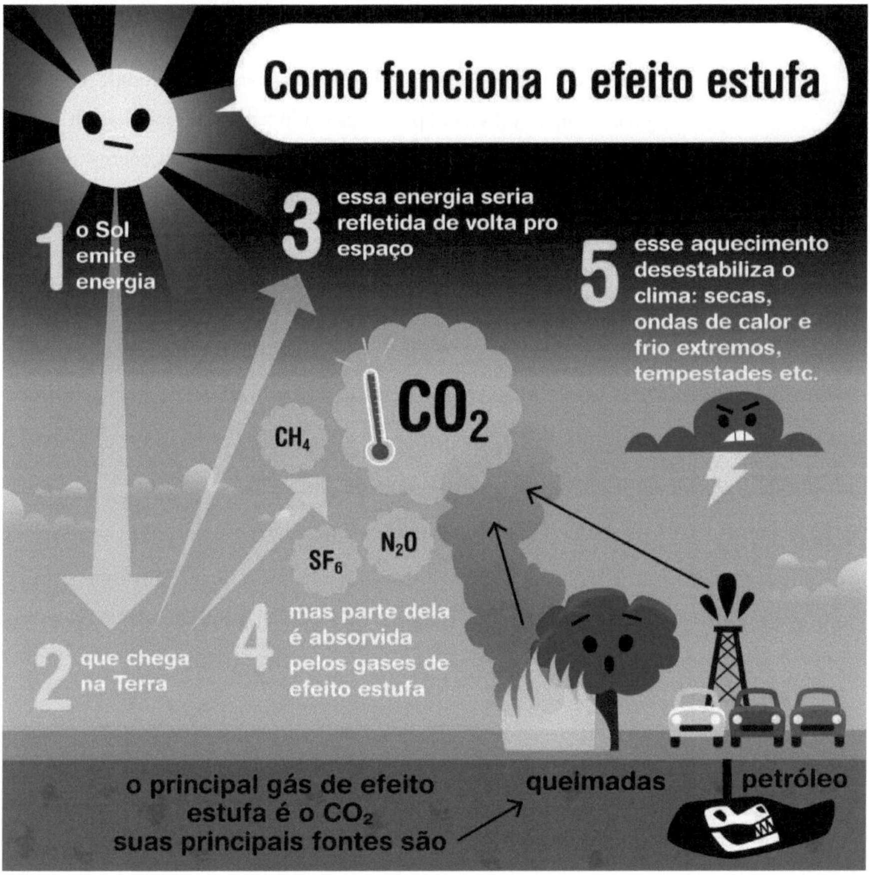

Source: Tree and water, 2023

The report also claims that "human-caused climate change is already affecting many weather and climate extremes in all regions of the world" (IPCC, p.60, 2023).

According to the report, one of the ways of adapting to mitigate global warming is the use of low-emission building materials, electrification in combination with low-emission sources, as well as technological innovation to reduce the emission of polluting gases and increase energy efficiency.

> "Mitigation interventions for buildings include: in the construction phase, low-emission building materials, highly efficient building envelope and the integration of renewable energy solutions; in the use phase, highly efficient appliances/equipment, the optimization of the use of buildings and their supply with low-emission energy sources; and in the disposal phase, recycling and reuse of building materials." (IPCC, p. 123, 2023)

Thanks to rising temperatures in the atmosphere, the demand for efficient and ecologically sustainable cooling systems is expected to increase. A self-sustaining material that cools an environment without high energy expenditure and the release of pollutants into the atmosphere, contributes to the reduction of greenhouse gases, as well as alleviating one of the effects of global warming (excessive heat) for its users. "Cities can achieve net zero emissions if emissions are reduced within and beyond their administrative boundaries through supply chains, creating beneficial cascading effects in other sectors." (IPCC, p. 123, 2023)

III. METHODOLOGY

A bibliographical review was used as the development method for this work. The material was researched on the internet by reading articles, finding information on websites, books and other tools that were useful for gathering information.

IV. PROTOTYPES TO REPLACE AIR CONDITIONING

A. HYDROCHERAMICS

"Hydroceramics" was the name given to the prototype of a material with the ability to insulate and cool the temperature of a closed environment without the need for any external assistance, such as the electricity grid or human handling. The material under development is a kind of ceramic that can change the temperature of an environment using only its own intelligence and depends solely on the climatic condition of the environment outside it.

> "The final prototype "Hydroceramics" works as an evaporative cooling device that reduces temperature and increases humidity and is capable of lowering the temperature of the indoor environment by around 5 to 6 degrees.Its built-in passive intelligence makes its performance directly proportional to the heat in the external environment, i.e. it cools more when it is warmer and does not cool when evaporation is not occurring (MITROFANOVA; RATHEE; SANTAYANON, 2013)."

The prototype, illustrated in Figure 2, was designed in 2013 by three students from the Institute of Advanced Architecture of Catalonia (IAAC): Akanksha Rathee, Elena Mitrofanova and Pongtida Santayanon. They were part of the "Digital Matter - Intelligent Constructions" research group, which aimed to improve efficiency in construction through digital simulations and manufacturing.

Figure 2 - Hydroceramics

Source: MITROFANOVA; RATHEE; SANTAYANON, 2013

Elena Mitrofanova, Akanksha Rathee and Pong Santayanon proposed a material with a dual function: to act as a thermal insulator and to cool closed spaces, promoting energy savings. The vision was to replace fans and air conditioning in homes and buildings. This material was created by combining hydrogel particles and ceramic plates made from ordinary clay.

> "Hydroceramics is a project that speculates on thermodynamic processes in buildings and how these can be addressed passively with a class of materials called "hydrogels". By combining the evaporation property of hydrogels with the thermal mass and moisture control property of ceramics and clay fabric, a composite material responsive to heat and water has been created. The proposed solution is a passive evapotranspiration system capable of lowering the temperature of an interior space by 5°C (INSTITUTE FOR ADVANCED ARCHITECTURE OF CATALONIA, 2014)."

Hydroceramics is a simple material made up of hydrogel, clay and laser-cut fabric. The hydrogel is the crucial element, as it defines the material's properties, but this will be detailed later. Clay is an organic material used to produce ceramics. During the testing phase, the clay is tested for thermal conduction and humidity. "Clay, Aluminum and Acrylic were tested against a control which helped determine that it is the porous nature of the clay that makes it help the cooling properties of the hydrogel in the best way." (MITROFANOVA; RATHEE; SANTAYANON, 2013)

The laser-cut fabric is a fabric with cut patterns and is mainly tested for water absorption and expansion. The focus of testing the components is to understand their conductivity and behavior in relation to water and expansion, principles related to the functionality of the hydrogel. Figure 3 below illustrates the hydrogel element.

Figure 3 - Hydrogel

Source: Institute for Advanced Architecture of Catalonia (IAAC), 2014

Hydrogel is a set of polymers that has the capacity to absorb up to 400 times its dry weight in a short time in contact with water. It also has a high capacity to retain and absorb water without the possibility of decomposition.

> "Chemically, they can be insoluble polymers of hydroxyethylacrylate, acrylamide, polyethylene oxide, among others. As a cooling aid, they work by exposing the absorbed water to a large surface area. As the heat of vaporization of water is around 0.6 kilocalories per gram, a cooling effect occurs (MITROFANOVA; RATHEE; SANTAYANON, 2013)."

With a few balls of hydrogel in a glass of water, the project's authors carried out a test to understand the hydrogel's absorption capacity on a smaller scale than that of a hydroceramic. The test resulted in almost total absorption within two and a half hours, as shown in Figure 4 below:

Figure 4 - Absorption capacity of the hydrogel

Source: MITROFANOVA; RATHEE; SANTAYANON, 2013

As the material requires few components, it is simple to assemble. The Hydroceramic is coated with a ceramic with bubble-shaped holes that help the hydrogel expand during growth without being damaged. Between the two ceramic plates is a laser-cut fabric that acts as a water channel, allowing water to flow through and come into contact with the hydrogel. This tissue also allows the hydrogel to expand. Finally, the hydrogel is inserted between the ceramic plate and the fabric, performing its main function in the material. Figure 5 below illustrates how the Hydroceramic is assembled.

Figure 5 - Composition and assembly of the Hydroceramics prototype

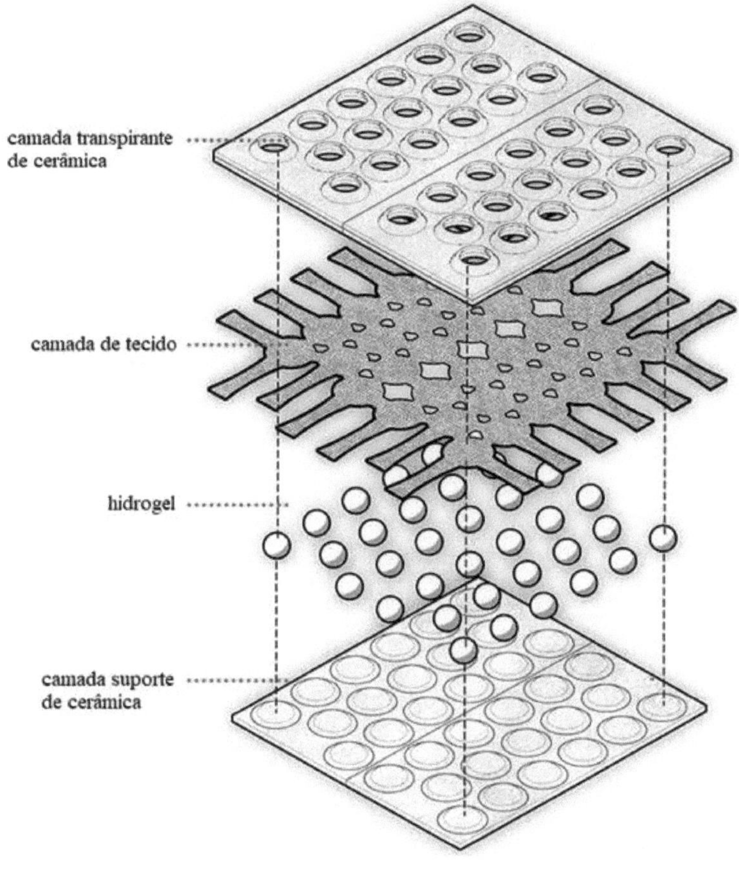

camada transpirante de cerâmica

camada de tecido

hidrogel

camada suporte de cerâmica

Source: MITROFANOVA; RATHEE; SANTAYANON, 2013, modified by the author

The expansion of the hydrogel when it comes into contact with water is the key mechanism for how hydroceramics work. On cold or rainy days (Figure 6), the hydrogel absorbs water from rain or humidity and increases in size. In a Hydroceramic plate, around 100 hydrogel balls are placed which, as they increase in size and weight, form an insulating layer that keeps the temperature of the room closed. It's worth noting that Hydroceramic doesn't "heat up" the room, but it does retain the room's temperature, generating thermal comfort on colder days.

Figure 6 - How the Hydroceramic Works on Rainy Days

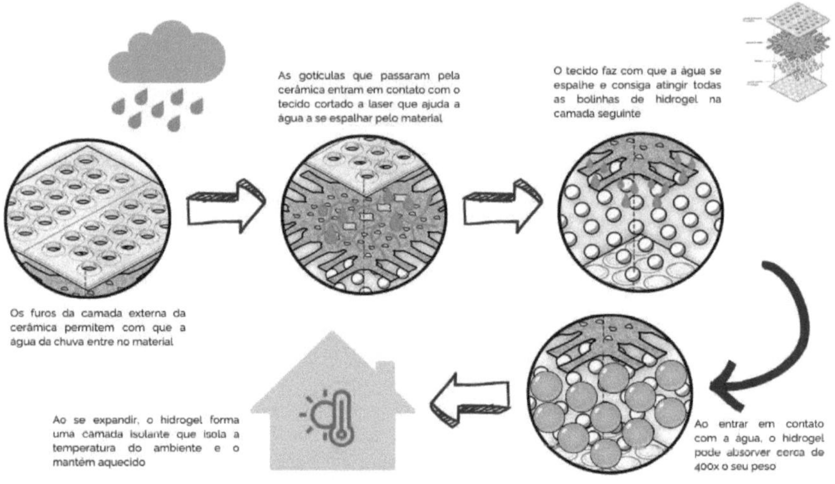

Source: Prepared by the author

On hot days (Figure 7), thanks to the heat, the water absorbed by the hydrogel evaporates. "A layer of fabric pulls in and channels the water, and behaves like a liquid transmitter throughout the system, while, due to its elasticity, holding the hydrogel in position" (MARKOPOULOU, p.13, 2019). The process releases water vapor and this vapor comes into contact with the clay (ceramic) which, in turn, will be responsible for lowering the temperature of the vapor and transferring cooling into the environment. This process is called evaporative cooling.

> "The clay layer towards the outside is designed and manufactured as a surface arranged with open conical shapes, to increase ventilation and allow access for water - in its various states - and air to reach the hydrogel. The conical geometries of this layer are customized based on the orientation of the building's façade to optimize both water absorption and the direction of evaporation (MARKOPOULOU, p.13, 2019)."

Figure 7 - How the Hydroceramic Works on Hot Days

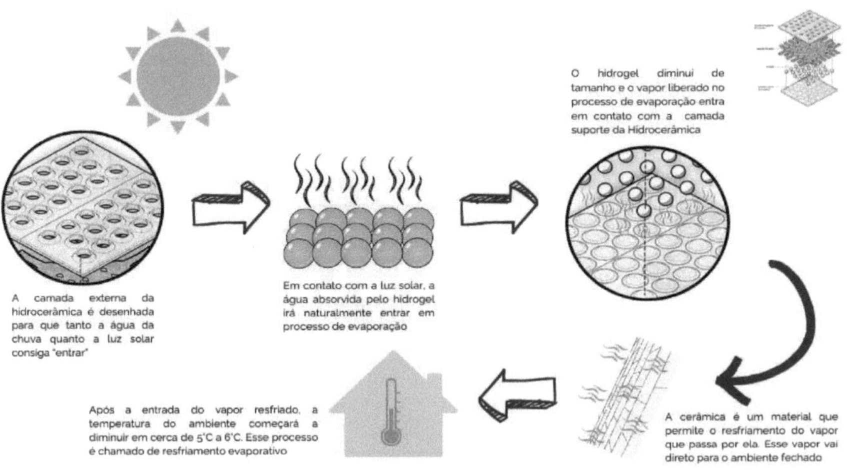

Source: Prepared by the author

According to PROCEL (2019) air conditioners are responsible for around 11% of all electricity consumed by household appliances in the country in the residential class, as shown in Figure 8. According to Mitrofanova, Rathee and Santayanon (2013), Hydroceramics "can help save up to 28% of the total electricity consumption caused by traditional air conditioning and can be used as a low-cost alternative construction technology, as both clay and hydrogel are inexpensive."

Figure 8 - Percentage Consumption of Electrical Appliances in Brazil

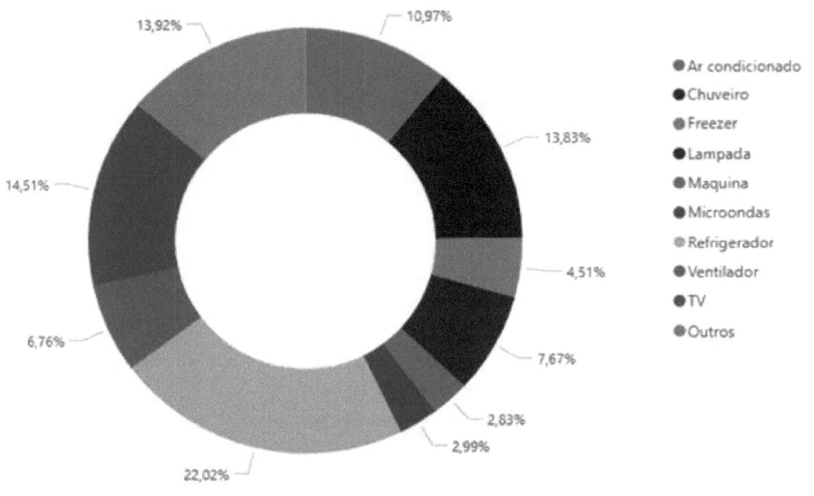

Source: PROCEL, 2019

As you can see, the material has the capacity to generate both environmental and economic impacts. According to a study carried out by Mitrofanova, Rathee and Santayanon (2014), an increase of just 1°C in air conditioning temperature results in a saving of 7% in total energy consumption. By using hydroceramics, it is possible to achieve an increase of 4°C in the appliance, which would result in a saving of 28% in energy consumption.

According to PROCEL (2019), the most commonly used air conditioners in the residential sector have a capacity of between 7051 and 10000 Btu/h. Figures 9 and 10 below show the most commonly used types of air conditioners and their thermal capacities, which will be used as the basis for the following calculations.

Figure 9 - Type of Air Conditioner most used in Brazil

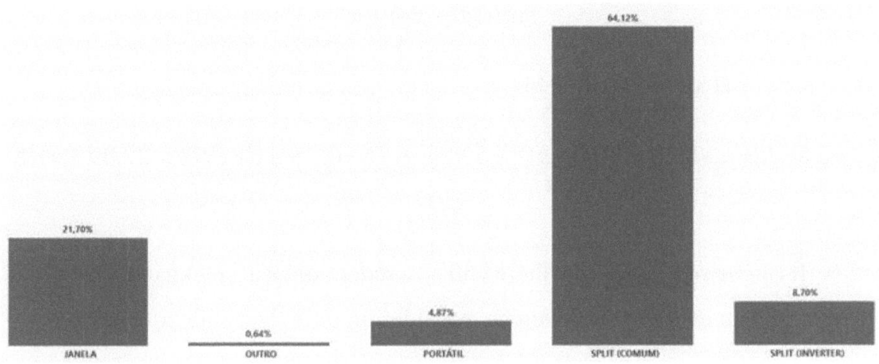

Source: PROCEL, 2019

Figure 10 - Thermal capacity of the most used appliances (BTU/h)

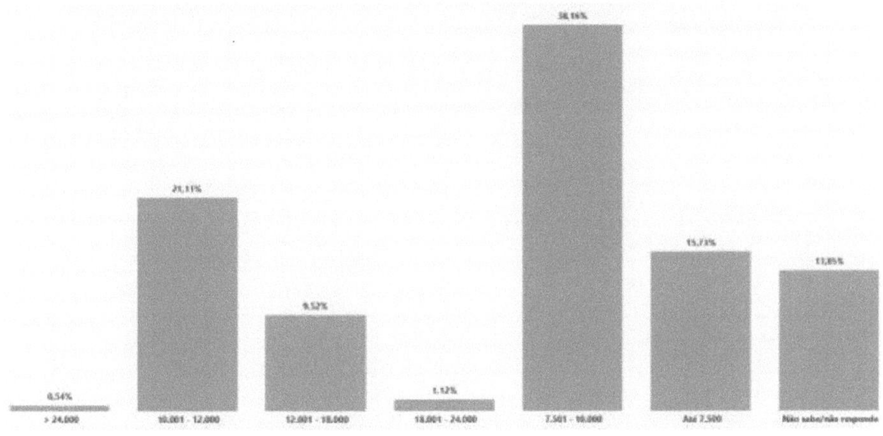

Source: PROCEL, 2019

Taking a split-type air conditioner with a capacity of less than or equal to 10,000 BTU/h as a base, which has an average consumption of 142.28 kWh per month (on for 8 hours a day) (Figure 11) and taking into account that every 1 kWh of electricity consumed generates approximately 0.7 kg of CO emissions$_2$, it can be calculated:

$$1\ kWh \rightarrow 0.7\ kg\ de\ \text{□□}2\ 142.28\ kWh * 0.7\ kg\ de\ \text{□□}2 = 99.60\ kg\ de\ \text{□□}2\ emitidos\ por\ \text{□ê□}$$

Hydroceramic helps save up to 28% of energy consumption:

$$142.28\ kWh * 0.28 = 39.84\ kWh\ 39.84\ kWh * 0.7\ kg\ de\ CO2 = 27.89\ kg\ de\ CO2$$

It can be concluded that there will be a reduction of 27.89 kg of CO_2 emitted monthly when using hydroceramics, considering only the residential case.

Figure 11 - Average Monthly Consumption of Electrical Appliances (KWh)

Aparelhos Elétricos	Dias Estimados Uso/Mês	Média Utilização/Dia	Consumo Médio Mensal (kWh)
Aparelho de blu ray	8	2 h	0,19
Aparelho de DVD	8	2 h	0,24
Aparelho de som	20	3 h	6,60
Aquecedor de ambiente	15	8 h	193,44
Aquecedor de mamadeira	30	15 min	0,75
Aquecedor de marmita	20	30 min	0,60
Ar-condicionado tipo janela menor ou igual a 9.000 BTU/h	30	8 h	128,80
Ar-condicionado tipo janela de 9.001 a 14.000 BTU/h	30	8 h	181,60
Ar-condicionado tipo janela maior que 14.000 BTU/h	30	8 h	374,00
Ar-condicionado tipo split menor ou igual a 10.000 BTU/h	30	8 h	142,28
Ar-condicionado tipo split de 10.001 a 15.000 BTU/h	30	8 h	193,76
Ar-condicionado tipo split de 15.001 a 20.000 BTU/h	30	8 h	293,68
Ar-condicionado tipo split de 20.001 a 30.000 BTU/h	30	8 h	439,20
Ar-condicionado tipo split maior que 30.000 BTU/h	30	8 h	679,20

Source: PROCEL, 2019

The hydroceramic material works automatically, without relying on energy or human intervention. In contrast, air conditioning requires energy and is affected by the "thermal load", which influences its consumption. According to Souza (2010), the thermal load can be described as the "rate of heat that must be removed from or supplied to an environment in order for it to maintain a constant temperature and humidity."

> "One of the factors responsible for the energy consumption of air-conditioning units is the thermal load, which depends directly on the design of the building. Thus, it can be seen that the thermal load due to insolation of a building contributes significantly to energy consumption, especially when air-conditioning units are used (SOUZA, p.15, 2010)."

As well as not consuming any energy, the material can be developed using a 3D printer. The creators of the prototype carried out their research considering the possibility of using this technology, instead of just preparing the material manually through the clay firing process. This new process allows the laser fabric to be discarded, as the water channels can be incorporated into the building, as well as reducing the cost and time of preparing the material.

The following figure (Figure 12) shows the structure of the Hydroceramic being manufactured using only the 3D printer.

Figure 12 - 3D printing

Explorar diferentes métodos de fabricação, como a impressão 3D, pode eliminar a necessidade de tecido, pois os canais de água podem ser incorporados nos edifícios e também reduzir custos, eliminando a necessidade de moldes.

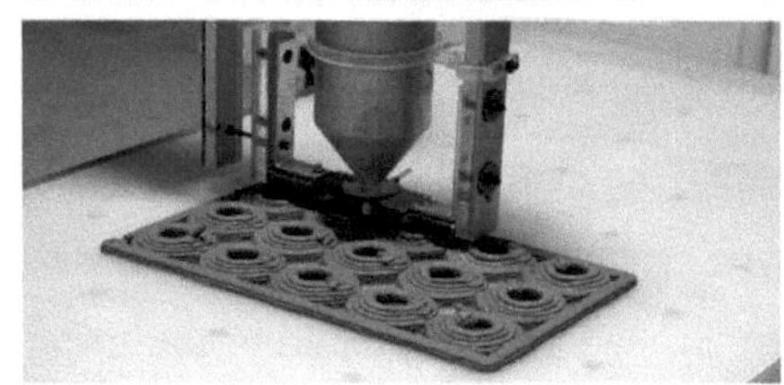

Source: MITROFANOVA; RATHEE; SANTAYANON, 2013, modified by the author

B. ENTREAUTRE CERAMIC REFRIGERATOR

In addition to the Hydroceramic, there are other prototypes that seek to reduce or replace air conditioning, with the aim of saving energy and reducing the environmental impact caused by the appliance. One example is the low-tech ceramic refrigerator (Figure 13), developed in 2022 by designers from the French studio Entreautre, which follows a similar approach to Hydroceramics and was created in ceramic with a 3D printer exclusively for the studio's laboratory, created by Dutch designer Olivier Van Herpét. "The machine is fed by a cylindrical tank into which the earth is placed and pushed by a piston into the nozzle. The earth is thus deposited layer after layer in a continuous wire, like a coil." (ENTREAUTRE, 2022)

Figure 13 - Low-tech ceramic refrigerator

Source: Entreautre, 2022

When the studio received the 3D printer, it used it to create ceramic objects that would be impossible to make by hand, and the models created ended up serving as inspiration to create the internal design of the ceramic refrigerator. According to Entreautre (2022), the prototype started from a "simple principle inspired by traditional practices: a (porous) terracotta container filled with water. Thanks to a ventilation system (REEE), the air flow in contact with the wet wall allows the water to evaporate to produce cold." Terracotta is a material composed of clay and water or baked clay; it has a reddish hue and is widely used in construction and the manufacture of ceramics. WEEE stands for Waste Electrical and Electronic Equipment, which Afonso, J. C. (2018) defines as:

> "[...] waste electrical and electronic equipment (WEEE) or simply electrical and electronic waste (WEEE), or even electrical and electronic waste (e-waste or WEEE), is equipment that has in its internal parts electrical and electronic components responsible for its operation, and which for reasons of obsolescence (perceptual or

> programmed) and impossibility of repair are discarded by their consumers (AFONSO, J. C., 2018)."

Examples of WEEE include cell phones, televisions, medical equipment and household appliances, among others. In the refrigerator, WEEE was used to create a ventilation system to increase the cooling effect of the air when it comes into contact with the ceramic material, as shown in Figure 14.

Figure 14 - Ventilation system integrated into the prototype (WEEE)

Source: Entreautre, 2022

The material has an interior structured in a similar way to a coral tree, forming a labyrinth that facilitates the circulation of water to humidify the walls, as shown in Figure 15. The porous nature of terracotta makes it possible to effectively absorb and retain moisture. When the moist material comes into contact with fluid air, the evaporation process begins, and the resulting fresh air is released into the environment, providing cooling in the room.

Figure 15 - Design of the Ceramic Cooler inside

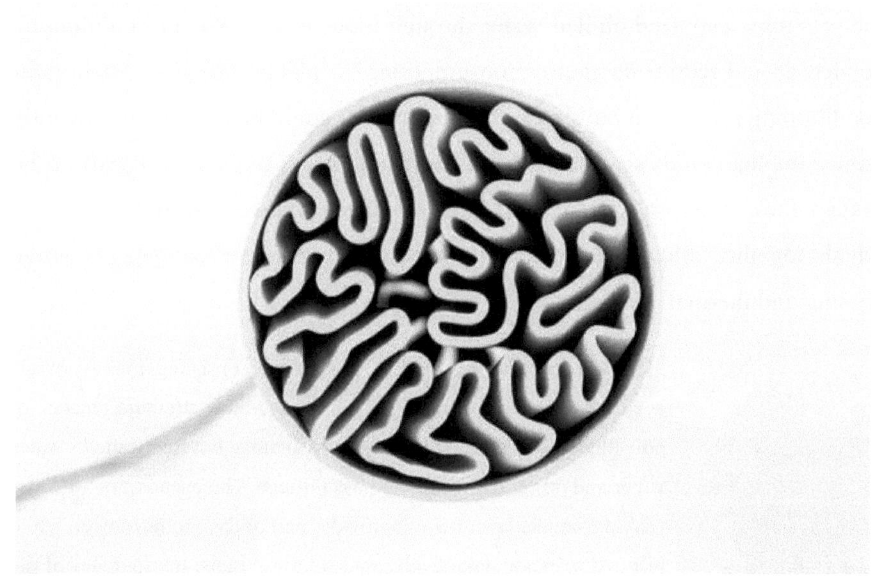

Source: Entreautre, 2022

C. MIT PROTOTYPE

Another prototype to be launched on the market was developed by the MIT (Massachusetts Institute of Technology) research team in 2022. This is a passive cooling device which, according to Chandler on the MIT portal (2022), aims to preserve food and complement conventional air conditioners in buildings, with no need for energy and only a small need for water.

> "The system, which combines radiative cooling, evaporative cooling and thermal insulation in a slim package that can resemble existing solar panels, can provide up to about 19 degrees Fahrenheit (9.3 degrees Celsius) of cooling from ambient temperature, enough to allow safe food storage for about 40 percent longer in very humid conditions. It can triple safe storage time in dry conditions (CHANDLER, 2022)."

The project aims to preserve food without high energy costs. In addition, the prototype can send chilled water through pipes to cool the air-conditioning condenser and reduce its energy consumption. "In places that have existing air conditioning systems in buildings, the new system could be used to significantly reduce the load on these systems by sending cold water to the hottest part of the system, the condenser." (CHANDLER, 2022) The system consists of three layers, which together allow the device to combine evaporative cooling, radiative cooling and thermal insulation.

> "The top layer is an aerogel, a material consisting mainly of air enclosed in the cavities of a sponge-like structure made of polyethylene. The material is highly insulating, but freely allows water vapor and infrared radiation to pass through. The evaporation of water (rising from the layer below) provides part of the cooling power, while infrared radiation, taking advantage of the extreme transparency of the Earth's atmosphere at these wavelengths, radiates some of the heat directly through the air and into space - unlike air conditioners, which expel hot air into the immediate surroundings (CHANDLER, 2022)."

Below the aerogel layer, there is a hydrogel layer that will be responsible for evaporative cooling and will work in the same way as in hydroceramics. According to Chandler (2022), below the hydrogel there is:

> "(...) a mirror-like layer reflects any incoming sunlight that has reached it, sending it back through the device instead of letting it heat the materials and thus reducing their thermal load. And the top layer of the aerogel, being a good insulator, is also highly reflective of solar air, limiting the amount of solar heating of the device, even in strong direct sunlight (CHANDLER, 2022)."

Unlike hydroceramics, the MIT cooler uses aerogel as a key component. However, this material is expensive, produced in a laboratory and requires special equipment to dry without damage. Its mass production requires further development. To solve the problem, according to Chandler (2022), the "research team is currently exploring ways to make this drying process cheaper, such as using freeze-drying, or finding alternative materials that can provide the same insulating function at a lower cost, such as membranes separated by an air gap." Unlike hydroceramics, the equipment requires maintenance as it needs water added every 4 days for the evaporation process to take place.

D. ADAPTIVE TILE

In 2023, scientists Charles Xiao, Bolin Liao and Elliot W. Hawkes from the University of California Santa Barbara in the USA developed an adaptive roof tile that also uses the concept of passive cooling and promises to reduce energy consumption in buildings and eliminate the need for air conditioning. It is a small prototype, just 10 cm^2, which acts as a passive thermoregulator and has the ability to change its thermal property over a temperature range.

> "We have built and tested a device that, depending on its temperature, switches automatically and passively (i.e. without electricity) between heating and cooling states. Compared to devices without switching, it reduces energy consumption for cooling by 3.1× and heating by 2.6×." (XIAO; LIAO; HAWKES, p.1, 2024, our translation)

According to the authors, around 50% of the energy consumed by buildings in the United States is used for heating and cooling. Thanks to climate change, this scenario could worsen and increase the demand for cooling and heating devices in buildings. "We can reduce the building's energy demand by reflecting sunlight and emitting infrared light into the space when it is hot and by absorbing sunlight and minimizing the emission of infrared light when it is cold (XIAO; LIAO; HAWKES, p.1, 2024)." Figure 16 below illustrates a graphic summary of how the adaptive roof tile prototype works.

Figure 16 - Graphical summary of the Adaptive Tile

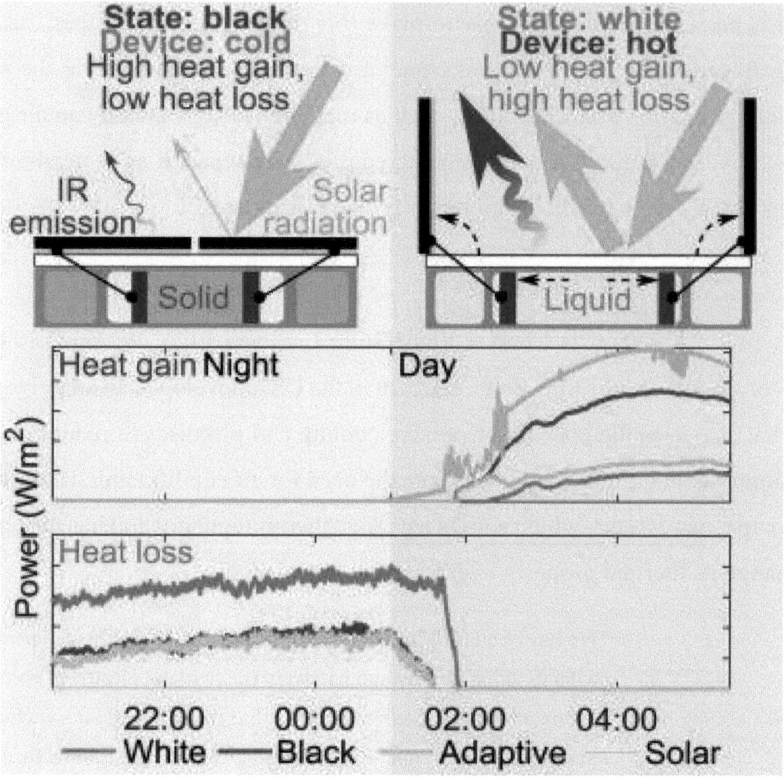

Source: XIAO; LIAO; HAWKES, 2024

The authors also explain that buildings may require temperature changes in the same day, i.e. they may need to be cooled during the day and heated at night, for example. Unfortunately, there are few technologies available that can switch between passive cooling and heating, and those that can, require a temperature range of more than 15°C for the cycle to change. The scientists claim that the prototype requires a range of just 3°C, with a target temperature of 18°C.

The device consists of a wax motor that doesn't require electronics or batteries and other materials that, according to the authors, can be made from cheap or recycled materials. Wax is classified as a phase-change material (PCM).

The wax motor is the actuating component that works by converting thermal energy into mechanical energy.

> "To achieve passive sensing and switching by opening and closing the shutters, we use a phase change material (PCM). These materials change from solid to liquid during heating, and have been used in building materials as a thermal capacitor to absorb large amounts of energy during temperature fluctuations (...) However, these materials also change volume during the phase change, which can be a disadvantage, especially for microencapsulation." (XIAO; LIAO; HAWKES, p.3, 2024, our translation)

The prototype's mechanism uses a wax motor and shutters that open and close according to the temperature and change in the state of the wax. When the shutters are closed, they have a black color that helps absorb solar radiation and when they are open, their interior has a white color that helps reflect solar radiation. The idea is that in cold weather, the wax used as a motor stays in its solid form and keeps the blinds closed, which will absorb solar radiation and emit little IR radiation, resulting in high heat gain. At higher temperatures, the wax expands and the blinds open, allowing solar radiation to be reflected and low heat gain.

> "When the temperature of the device is above the setpoint, the PCM expands and drives the shutters open, resulting in low heat gain. Solar radiation is reflected, and large amounts of IR radiation are emitted into deep space." (XIAO; LIAO; HAWKES, p.2, 2024, our translation)

Figure 17 below illustrates how the adaptive tile works in more detail.

Figure 17 - How the Adaptive Tile works

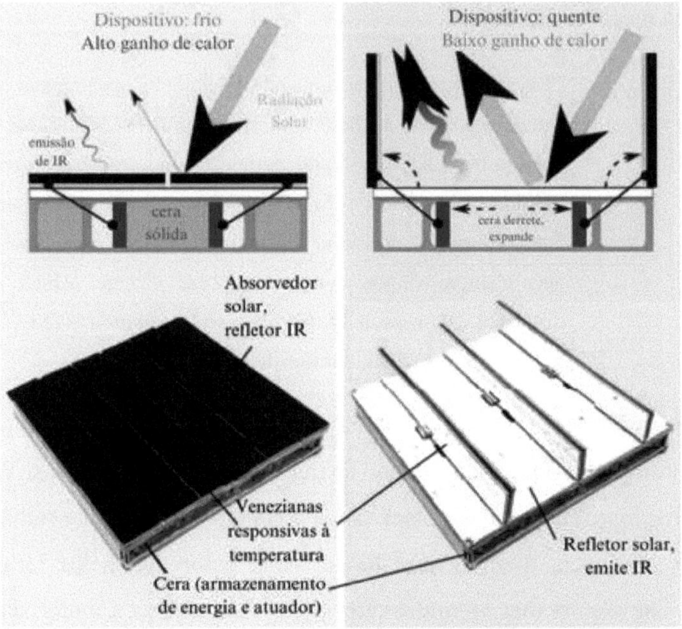

Source: XIAO; LIAO; HAWKES, 2024, modified by the author

Charles Xiao, Bolin Liao and Elliot W. Hawkes conclude the project by pointing out the high adaptability of the device created, thanks to the minimal restriction of materials compatible with it.

> "Different temperature set points can be achieved simply by using different phase change materials and higher radiative cooling and heating performance or high levels of durability can be achieved by different coatings" (XIAO; LIAO; HAWKES, p.2, 2024).

The authors also suggest for future work a solution for the adaptability and longevity of moving parts for the device. They also suggest a solution for the effective cooling of the blinds, as in their testing phase there is a higher than ideal temperature rise in temperature tests and heat gain in the power test. According to Xiao, Liao and Hawkes (2024), "future studies need to be carried out on larger arrays to minimize boundary effects and in the development of improved louver cooling methods."

E. NATURAL AIR CONDITIONING - ANT STUDIO

In India, the architecture studio Ant Studio has developed an intelligent and sustainable solution based on evaporative cooling, similar to the principle used by the scientists who created hydroceramics. They have created an external cooling system that uses inexpensive materials such as clay, recycled water and reusable steel, while requiring little electricity.

The system was installed at the Deki Electronics factory in Noida, next to the generator set (Figure 18), which was placed at the entrance to the factory, on the sidewalk. This exposed the workers to an intense flow of heat every day. Heat waves are common in the country, especially in summer, with temperatures that can reach 40°C. Added to the heat from the generators, this created a situation that was detrimental to the workers' health. "The temperature of the air flow around the installation was recorded. It was noticed that the hot air entering the installation was above 50 degrees Celsius at a speed of 10m/sec." (ANT STUDIO, S.D.)

Figure 18 - Ant Studio cooling system

Source: S. Anirudh[1]

As a solution, a team from the studio made up of Monish Siripurapu, Abhishek Sonar, Atul Sekhar and Sudhanshu Kumar proposed an economical cooling system that would meet the air outlet of the generators to cool the temperature and reduce the heat flow.

The system's main material is terracotta, the same used by Entreautre. It is made up of water and clay or baked clay, the porous nature of which makes it possible to effectively absorb and retain moisture. The principle is similar to the others presented: when the moist material comes into contact with fluid air, the

[1] Photographer responsible for the photos of the Ant Studio project. Available at: <https://ant.studio/beehive/qy4z4lq8uradkygqlbsj43lhbughe1>

evaporation process begins and the resulting fresh air is released, providing cooling. The use of cylindrical cones increases the surface area and maximizes the effect.

Recycled water from the factory is used to run over the surface of the clay cones and cool the hot air from the generators (Figure 19). It is more economical and sustainable, but periodic maintenance is required to clean the pores of the cones. By using non-recycled water, the system's performance changes over the long term, but it doesn't prevent the need for less frequent cleaning.

Figure 19 - Recycled water cooling system

Source: S. Anirudh

According to Ant Studio (n.d.), "it was observed that, after achieving the cooling effect, the temperature around the set up dropped to 36 degrees Celsius, while the outside temperature remained high at 42 degrees Celsius. And the air flow was recorded as 4m/sec."

The design was inspired by a beehive, and the idea was for the system to be both effective and visually appealing, as if it were art. Several clay cones were used to maximize the cooling effect.

> "The solution is ecological and artistic, and at the same time involves traditional craft methods. It allows for a low-maintenance, sustainable and inexpensive alternative using porous terracotta and its inherent cooling properties, converting the hot air from the generator set into a pleasant breeze." (ANT STUDIO, S.D.)

The studio claims that the cooling system is ecological, economical and aesthetic. It's ecological because the materials used, such as clay, are fully recyclable and can be returned to nature. It is economical because it uses cheap and easily available local materials, as well as helping the factory to reduce its energy consumption, requiring little electricity just for the water pump that circulates the water recycled by the system repeatedly. With regard to aesthetics, Ant Stdio (n.d.) states:

> "The idea is to break the usual rule of keeping generators in the site's rear area. The idea is to move them from the backyard to the front yard. In India, generator systems are usually placed in setback areas where loading and unloading takes place. We reflected very intrinsically on the symmetrical geometry of the beehive and designed our product so that it can be interpreted as an art installation rather than just playing the functional role of an air conditioner." (ANT STUDIO, S.D.)

For the project, the studio has the prospect of using it as an element of mass production globally. Many factories in India are experiencing similar problems and the prototype can be used as a solution. What's more, the company announces that the system can require zero energy if it is powered manually once or twice a day. There are also prospects for using the product indoors.

It's interesting to add that Ant Studio has a solution called Cool Ant. It works in conjunction with architecture, but is motivated by the purpose of helping buildings to "fight the heat"; it is an initiative that tries to reduce the effects of climate change by using natural solutions, combining tradition and technology, mixing art and science and without using air conditioning.

Resolution aims to reduce the carbon footprint (CO_2) and to do this, the best way is to design the building from the outset to have passive cooling. According to the studio, most of the heat in buildings comes from the roof and facades, so one of their solutions is to create solar protection barriers on these surfaces, as if it were a "second skin" for the building. This is what they call the AeroLeaf Facade.

According to research carried out by the studio itself, this façade can save around 30% of energy based on passive cooling and can reduce the temperature of a room by up to 15%. Figure 20 below shows an element of this façade.

Figure 20 - AeroLeaf façade

Source: S. Anirudh

There are other materials developed by the studio that contribute to sustainable cooling, using terracotta and passive cooling as the basic principles of each project.

V. CONCLUSION

The construction sector has the potential to integrate technology and deal with its environmental impact. Hydroceramics offers an intelligent cooling and thermal insulation system that reduces the need for air conditioning, cutting energy consumption and carbon dioxide emissions, as well as being an automatic and low-cost system. The Entreautre project uses porous materials such as terracotta and explores innovative technologies such as 3D printing. The combination of technologies, as demonstrated by the Massachusetts Institute of Technology (MIT) prototype, has the potential to create more effective materials, even considering cost and complexity factors. The study conducted by scientists at the University of California Santa Barbara, resulted in the creation of an innovative adaptive roof tile that uses passive cooling to reduce energy consumption in buildings, using the properties of wax for its operation. Ant Studio developed an innovative and sustainable solution for evaporative cooling, addressing the problem of extreme heat faced by workers at the Deki Electronics factory, and using cheap and recyclable materials such as clay, recycled water and reusable steel.

These sustainable cooling solutions encompass intelligent, multidisciplinary systems that aim for a self-sustainable future with less energy expenditure and carbon dioxide emissions, factors that are extremely important for climate issues, such as helping to reduce the effects of global warming.

Although prototypes and techniques help to reduce the use of air conditioning and, in a way, reduce environmental impacts, they are far from the solution to the problem. The large-scale reduction of global warming must be explored where its greatest cause lies: industries and the agricultural sector.

As technology advances, new materials will be developed, using an ever-increasing variety of disciplines and technologies. This will boost the construction of homes and buildings with sustainable cooling systems and will allow the creation of chillers made with cheaper and more environmentally friendly materials, reducing their final cost. In addition, the techniques presented in this study are more accessible and help to cool environments efficiently, benefiting those who cannot afford air conditioning.

The literature review presented in this material serves as a basis and incentive for future work on the development of these prototypes or new materials and similar techniques.

VI. REFERENCES

AFONSO, Júlio Carlos. Waste Electrical and Electronic Equipment: the anthropocene knocks on our door. **Revista Virtual de Química**, [S.L.], v. 10, n. 6, p. 1849-1897, 2018. Brazilian Chemical Society (SBQ). http://dx.doi.org/10.21577/1984-6835.20180121.

ANT STUDIO (India). **DESIGN AND INSTALLATION OF AN ECONOMICAL AND ECOLOGICAL EXTERNAL COOLING SYSTEM**. Available at: <https://ant.studio/beehive/mtqby6ivdkyudroyknv708m0ptyhco>. Accessed on: June 10, 2024.

BRAZILIAN ASSOCIATION OF PUBLIC CLEANING AND SPECIAL WASTE COMPANIES (ABRELPE). **Panorama of Solid Waste in Brazil 2022**. São Paulo: ABRELPE, 2022. Available at: < https://abrelpe.org.br/panorama/>. Accessed on: October 2, 2023.

BRAZIL. Law n. 10.973, of December 2, 2004. Provides incentives for innovation and scientific and technological research in the productive environment and makes other provisions. **Federal Official Gazette**, Brasília, DF, December 3, 2004.

CHANDLER, D. L.. **Passive cooling system could benefit off-grid locations**. 2022. Available at: <https://news.mit.edu/2022/passive-cooling-off-grid-0920>. Accessed on July 30, 2023.

CONCEIÇÃO, J. F. da; SANTOS, M. P. dos . Sustainable Construction. **Epitaya E-books**, *[S. l.]*, v. 1, n. 6, p. 426-458, 2021. DOI: 10.47879/ed.ep.2021250p426. Available at: <https://portal.epitaya.com.br/index.php/ebooks/article/view/194>. Accessed on June 28, 2023.

greenhouse effect. 2023. Available at: https://arvoreagua.org/crise-climatica/efeito-estufa. Accessed on: 08 May 2024.

ENTREAUTRE (France). **Our experiences around ceramic 3D printing**. 2022. Available at: <https://www.entreautre.com/case-study/i3d-ceramique-design-recherche-drome/>. Accessed on July 29, 2023.

FERNANDES, A. V. B.; AMORIM, J. R. R. Sustainable concrete applied in civil construction. Cadernos de Graduação Ciências Exatas e Tecnológicas, Aracaju, v. 2, n. 1, p.79-104, jun. 2014. Available at <https://periodicos.set.edu.br/index.php/cadernoexatas/article/view/1093>. Accessed on: April 6, 2023.

INSTITUTE FOR ADVANCED ARCHITECTURE OF CATALONIA (IAAC). **Hydroceramics**, 2014. Available at: <https://iaac.net/project/hydroceramic/>. Accessed on 15 Mar. 2023

IPCC, 2023: Sections. In: Climate Change 2023: Synthesis Report. Contribution of Working Groups I, II and III to the Sixth Assessment Report of the Intergovernmental Panel on Climate Change [Core Writing Team, H. Lee and J. Romero (eds.)]. IPCC, Geneva, Switzerland, pp. 35-115, doi: 10.59327/IPCC/AR6-9789291691647

MARKOPOULOU, Areti. **DESIGN BEHAVIORS**: programming the material world for responsive architecture. 2019. 160 f. Thesis (Doctorate) - Degree in Architecture, Department: Proyectos Arquitectónicos, Universidad Politécnica de Catalunya (Upc), Barcelona, 2019.

MITROFANOVA, Elena; RATHEE, Akanksha; SANTAYANON, Pong. **HYDROCERAMIC**. Materiability Research Group, 2013. Available at: <http://materiability.com/portfolio/hydroceramic/>. Accessed on September 13, 2022.

POTT, Luana Mariana; EICH, Monique Costa; ROJAS, Fernando Cuenca. Technological innovations in civil construction. **Interinstitutional Seminar on Teaching, Research and Extension**, v. 22, 2017.

PROCEL. **Energy Saving Tips**. Disponível em: <http://www.procelinfo.com.br/main.asp?View=%7BE6BC2A5F-E787-48AF-B485-439862B17000%7D>. Accessed on May 21, 2023.

RATHEE, Akanksha; MITROFANOVA, Elena; SANTAYANON, Pongtida. **HYDROCERAMIC**. 2013. 55 f. Dissertation (Master) - Architecture Course, Institute For Advanced Architecture Of Catalonia, Barcelona, 2013. Available at: https://issuu.com/akanksharathi/docs/final_booklet. Accessed on: June 2, 2024.

ROTH, C. das G.; GARCIAS, C. M. Civil Construction and Environmental Degradation. **Desenvolvimento em Questão**, *[S. l.]*, v. 7, n. 13, p. 111-128, 2011. DOI: 10.21527/2237-6453.2009.13.111-128. Available at: <https://www.revistas.unijui.edu.br/index.php/desenvolvimentoemquestao/article/view/169>. Accessed on April 19, 2023.

Denise Scabin. Environmental Education Portal. **GLOBAL WARMING**. 2023. Available at: https://semil.sp.gov.br/educacaoambiental/prateleira-ambiental/aquecimento-global/#:~:text=O%20aquecimento%20global%20%C3%A9%20o,)%20e%20vapor%20d%27%C3%A1gua.. Accessed on: 30 Apr. 2024.

SOUZA, Edson Palhares de. **ENERGY SAVING IN AIR CONDITIONING IN BRAZIL**: efficiency and economy. 2010. 116 f. Dissertation (Master's Degree) - Energy Engineering Course, Federal University of Itajubá, Itajubá, 2010. Available at: <https://repositorio.unifei.edu.br/jspui/bitstream/123456789/1451/1/dissertacao_0036194.pdf>. Accessed on 05 Apr. 2023.

UNITED NATIONS ENVIRONMENT PROGRAM. **2022 Global status report for buildings and construction**: towards a zero-emissions, efficient and resilient buildings and constructions sector. Nairobi, 2022.

XIAO, Charles; LIAO, Bolin; HAWKES, Elliot W.. Passively adaptive radiative switch for thermoregulation in buildings. **Device**, [S.L.], v. 2, n. 1, p. 100186, jan. 2024. Elsevier BV. http://dx.doi.org/10.1016/j.device.2023.100186. Disponível em: <https://www.cell.com/device/fulltext/S2666-9986(23)00304-6?_returnURL=https%3A%2F%2Flinkinghub.elsevier.com%2Fretrieve%2Fpii%2FS2666998623003046%3Fshowall%3Dtrue>. Accessed on: 26 Jan. 2024.

Printed by Books on Demand GmbH, Norderstedt / Germany